THE WATER CYCLE™

WATER IN GLACIERS

Isaac Nadeau

The Rosen Publishing Group's
PowerKids Press™
New York

To Garrett

Published in 2003 by The Rosen Publishing Group, Inc.
29 East 21st Street, New York, NY 10010

First Edition

Editor: Gillian Houghton
Book Design: Maria E. Melendez

Photo Credits: Cover, title page, page borders, p. 20 (Earth) © Digital Vision; pp. 4, 7, 8, 11, 12, 15, 16, 19, 20 (illustrations) by Maria E. Melendez.

Nadeau, Isaac.
Water in glaciers / Isaac Nadeau.
p. cm. — (The Water cycle)
Includes bibliographical references (p.).
ISBN 0-8239-6265-2 (lib. bdg.)
1. Glaciers—Juvenile literature. 2. Hydrologic cycle—Juvenile literature. [1. Glaciers. 2. Hydrologic cycle.] I. Title.
GB2403.8 .N34 2003
551.31'2—dc21
2001006172

Manufactured in the United States of America

CONTENTS

atmosphere
ocean
stream

THE WATER CYCLE

Water on Earth's surface, under ground, and in the **atmosphere** takes on many forms. Water can be a liquid as in streams, lakes, and oceans. It can be a gas in the form of water vapor floating in the air. Water also can be a solid, frozen as in a glacier. Water is carried from one place to another in a process called the water cycle. There is the same amount of water on Earth today as there has been for millions of years. Water does not disappear or get used up. It changes from one form to another as it moves through the water cycle. A drop of water might fall to Earth as rain or snow. Over time it might be part of the ocean, soak deep into the ground, float high above in a cloud, or become part of a glacier.

Water in glaciers or ice sheets might stay frozen for thousands of years, but eventually even this water moves on to another part of the water cycle.

THE ICE AGE

Beginning about 2.4 million years ago and lasting until about 10,000 years ago, Earth experienced a time of colder temperatures. This period is known as the **Pleistocene epoch**. As the planet's temperature cooled, ice formed in many places around the planet. For many thousands of years at a time, huge glaciers called ice sheets advanced over the land. They formed in the cold areas near the poles and extended toward the **equator**. These times are known as ice ages. When Earth warmed slightly, the ice sheets melted and retreated. These periods between ice ages are known as **interglacial periods**. There were several ice ages and interglacial periods during the Pleistocene epoch. The most recent ice age began about 100,000 years ago and ended 10,000 years ago.

During the last ice age, known as the Wisconsin Glaciation, glaciers advanced over present-day Canada and the midwestern United States (inset). During this period, Earth was on average 10°F (6°C) colder than it is today. Ice sheets covered 30 percent of Earth's land surface.

2.4 million years ago
250,000 years ago
100,000 years ago
10,000 years ago
Present

Pleistocene epoch

The first modern humans, Homo sapiens, walk Earth.

The most recent ice age begins.

The modern epoch, known as the Holocene epoch, begins.

Most glaciers can be divided into two zones, or areas. The top of a glacier, called the zone of accumulation, is usually the coldest and receives the most snow. The bottom of the glacier is usually warmer. There ice often melts and drains away. As snow accumulates, the frozen water changes shape under its combined weight, from flake to firn to glacier ice (above).

HOW GLACIERS FORM

Glaciers can be formed wherever the yearly snowfall is greater than the amount of snow that melts each year. Glaciers form in cold places, such as high in the mountains or in polar regions. Unmelted snow accumulates, or collects, on the ground. A snowflake is made up mostly of air. As the air is squeezed out of them, snowflakes are pressed tightly together. At first snowflakes form small, tight balls of ice called **firn**. After many years, this ice becomes **compressed** under its own weight. As more snow accumulates and weight is added above, the balls of firn are pressed together, forming larger and larger crystals known as glacier ice. Eventually the glacier ice grows large enough that it begins to move slowly downhill. It is pulled by the force of **gravity**. When this happens, the ice is known as a glacier.

ICE SHEETS AND ICE CAPS

Not all glaciers are alike. Glaciers are given different names depending on their size, their shape, or their location. The largest glaciers are called ice sheets. The Antarctic ice sheet is twice as large as the entire continent of Australia and is up to 3 miles (5 km) thick. There is far more freshwater frozen in this ice sheet than in all of the lakes, streams, and rivers on Earth combined. Another large ice sheet covers most of Greenland. This ice sheet is about the size of the country of Mexico. Both the Antarctic and Greenland ice sheets are in polar regions, where the temperatures remain cold all year. Ice caps are second in size to ice sheets, but they are still much larger than most glaciers. Ice caps also are common in polar regions.

ANTARCTICA

nunataks

Ice caps and ice sheets, such as the Antarctic ice sheet, are so large that entire mountain ranges can be buried beneath them. Sometimes just the peaks of mountains, called nunataks, can be seen sticking out above the ice. Nunatak *means "island in a sea of ice" in* Inuit*, a language spoken by a people native to the Arctic.*

cirque

iceberg

valley glacier

Valley glaciers (above) *can advance from the mountains to the sea, where they become tidewater glaciers. Ice often breaks off of the front end of tidewater glaciers, producing icebergs that float out to sea* (top right). *A cirque glacier advances through a valley, carving a steep-walled basin, or bowl, in the mountainside* (top left).

SMALLER GLACIERS

Smaller glaciers can be found at the poles, as well as in the mountains of **temperate** regions. A valley glacier can be found advancing from a large mountain ice cap. A valley glacier can be as small as several thousand feet (m) or more than 200 miles (322 km) long. Many valley glaciers move down valleys that were originally cut by streams. A **cirque** glacier carves a bowl-shaped hollow, or cirque, in the side of a mountain. As the glacier ice moves, it cuts the cirque deeper. Most cirque glaciers were once part of larger glaciers, but broke off as the ice melted. A hanging glacier is a glacier that forms on a steep mountainside. As snow accumulates on top, gravity pulls with greater force. This often causes the ice at the bottom of hanging glaciers to break off, sending an **avalanche** of snow, ice, and rock to the valley below.

HOW GLACIERS MOVE ON LAND

All glaciers travel over land. There are two main ways that a glacier moves. The first, called creep, occurs when the ice inside the glacier changes shape because of the great weight of the ice above it. The glacier moves as bread dough does on a tilted baking sheet. Sometimes as the glacier moves along, it splits, causing a crevasse, or crack, in the ice that can be as much as 100 feet (30 m) deep. The second way a glacier moves is called basal sliding. The weight of the glacier lowers the point at which water freezes at the bottom of the glacier. This process causes its base, or underside, to melt slightly. The glacier is able to slide slowly over the land. Glaciers move at different speeds depending on their size, the angle of the mountain or hillside, the amount of water melting from the glaciers, and the objects that block their way.

Glaciers advance and retreat depending on how much snow they accumulate or lose and on changes in temperature. Ice sheets and ice caps advance and retreat in all directions from their edges. Valley glaciers advance and retreat from their snouts, the part of the glacier farthest downhill. A valley glacier slides down a mountainside (1), through a valley (2), and across a plain (3). Most glaciers only move a few feet (m) per year, but some glaciers have moved as much as 130 feet (40 m) in a single day.

mountain lake

As glaciers advance and retreat, they carve the land. A mountain lake (above) is formed by the collection of meltwater in a hollow carved by a glacier.

SHAPING THE LAND

As glaciers move, they change the landscape beneath and around them. Glaciers contain more than ice. They contain pieces of earth of all shapes and sizes, from grains of sand to car-sized boulders. As a glacier moves down a valley or along the sides of mountains, these rocks scrape against the surface of Earth. The rocks carve out valleys and smooth over hills. Glaciers erode, or wear away, the land over which they travel. When cirque glaciers melt, bowl-shaped cirques are left. These cirques fill with melting snow or rainwater and form mountain lakes called **tarns**. A **kettle** forms where a large piece of ice breaks off as an ice sheet retreats. This ice is buried under the sand and rock left behind by the melting glacier. As this buried ice melts over hundreds of years, it leaves behind a kettle-shaped bowl in the ground.

GLACIAL FORMATIONS

Glaciers pick up rocks as they advance. When a glacier retreats, these rocks are left behind. They range in size from grains of sand to giant boulders. These pieces of earth are known as **till**. As the snout of an advancing glacier moves across the land, till is pushed out of the way, swept up and carried along, or hidden under the ice. As a glacier retreats, it often leaves behind ridges of till that had collected at its snout for many years. These hills are called **moraines**.

Streams of melting glacier ice carve tunnels in the ice just above the ground. This melted water carries till with it. Over time the tunnels fill with till and become solid ground. When the glacier melts and retreats, long curving hills of till called **eskers** are left behind. Eskers trace the paths of the streams that once flowed beneath the glacier.

moraine

As a glacier melts, a stream of meltwater and till flows from its snout. Where glaciers once existed or exist today, a valley might be filled with moraines formed by this glacial stream. These gravel-filled valleys are called outwash plains.

Heat from the Sun passes through the atmosphere. Most of it is absorbed by Earth. The rest is reflected back into the atmosphere. Earth gives off a lot of heat, too. Much of this heat is reflected back to Earth by the natural gases in the atmosphere, just as a gardener's greenhouse traps heat within its glass or plastic walls (inset). This heat is necessary to keep Earth warm. However, when there is too much gas in the atmosphere, not enough heat can escape.

GLACIERS AND GLOBAL WARMING

Earth is kept warm by the greenhouse effect. Gases in the atmosphere, such as **carbon dioxide**, help trap heat from the Sun close to Earth's surface. However, humans have increased the amount of carbon dioxide in the atmosphere. Cars and factories release the gas into the air. This increases the amount of heat trapped near Earth's surface and causes average yearly temperatures to rise slightly. This process is called global warming. Many scientists predict that glaciers around the world will continue to melt. As glacier ice melts, it flows into the surrounding oceans and causes them to rise slowly. Changes in sea level affect the weather and the lives of the people who live close to the ocean. Most scientists agree that it is important to study the relationship between global warming and the melting of glaciers.

GLACIERS TODAY AND TOMORROW

Scientists who study glaciers are called glaciologists. Many glaciologists believe that we are living in an interglacial period. In the future, Earth will cool. Great ice sheets will advance across the land. Plants and animals that cannot live in the cold or move to warmer places will die. Farms will be buried under ice. It will be hard to find places to fish or hunt. However, humans have lived during ice ages before and would be likely to live through another. Even today glaciers are part of the water cycle. Glaciers cover 10 percent of Earth's land surface, and all of the glaciers described in this book can be found on Earth today. By studying today's glaciers, we can learn about the glaciers of the past and see something of what the future holds.

GLOSSARY

atmosphere (AT-muh-sfeer) The layer of air that surrounds Earth.
avalanche (A-vuh-lanch) Snow, ice, rock, or mud that falls suddenly down a mountainside.
carbon dioxide (KAR-bin dy-OK-syd) A gas that is found in the atmosphere and is important to most living things.
cirque (SURK) A bowl-shaped hollow in a mountainside, carved by a cirque glacier.
compressed (kum-PREST) Squeezed something into a smaller place.
equator (ih-KWAY-tur) An imaginary line around Earth that separates it into two parts, North and South.
eskers (ES-kurz) Long, winding hills formed by deposited till.
firn (FIRN) Tightly packed balls of ice.
gravity (GRA-vih-tee) The natural force that causes objects to move toward the center of Earth.
interglacial periods (in-tuhr-GLAY-shul PEER-ee-uhdz) Periods of warm temperatures between ice ages.
kettle (KEH-tul) A bowl-shaped hollow in the ground.
moraines (muh-RAYNZ) Hills of deposited till.
Pleistocene epoch (PLYS-tuh-seen EH-pihk) The period of time between 2.4 million years ago and 10,000 years ago.
tarns (TAHRNZ) Mountain lakes formed in cirques.
temperate (TEM-peh-ret) Not too hot and not too cold.
till (TIHL) Rocks of all sizes, deposited by glaciers.

INDEX

WEB SITES

To learn more about glaciers, check out these Web sites:
http://nsidc.org/glaciers/story/page1.html
www.glacier.rice.edu/